Independent Schools
Examinations Board

MATHS
WORKBOOK FOR COMMON ENTRANCE
13+

KEEPING UP WITH THE JONESES

BARBARA LANGFORD

A John Catt Publication

First Published 2016

by John Catt Educational Ltd,
12 Deben Mill Business Centre, Old Maltings Approach,
Melton, Woodbridge IP12 1BL
Tel: +44 (0) 1394 389850 Fax: +44 (0) 1394 386893
Email: enquiries@johncatt.com
Website: www.johncatt.com

ISBN: 9781-911382-08-9

Set and designed by Theoria Design Limited
www.theoriadesign.com

Printed and bound in Great Britain

INTRODUCTION

CHAPTER 1

The Party (ratio, significant figures and multiplication)

CHAPTER 2

In the Kitchen (volume, nets)

CHAPTER 3

In the Garden (area, scale drawing, volume, ratio and Pythagoras' theorem)

CHAPTER 4

The Journey to School (primes, multiples and factors, speed, distance, time)

CHAPTER 5

At School (fractions, percentages, probability, speed and distance)

CHAPTER 6

The Day Trip (addition and subtraction, tessellations,
angles in polygons and transformations, percentages)

CHAPTER 7

The Skiing Holiday (conversions, percentage increase, ratio)

CHAPTER 8

The Sailing Holiday (bearings)

CHAPTER 9

The French Gite (percentages, wordy fractions, area and
volume with circles, simultaneous equations)

CHAPTER 10

Back at home relaxing playing games
(Algebra – factorisation and brackets, equations, substitution)

INTRODUCTION

KEEPING UP WITH THE JONESES - A COMMON ENTRANCE (13+) WORKBOOK

This book is written as a lighthearted way to supplement and practise predominantly level 2 Common Entrance 13+ work. However, the challenge questions are level 3 or CASE (Common Academic Scholarship Entrance) level.

If you are looking to recap and boost maths learning at home, with any Year 8 pupil, then this should be a suitable level for Intermediate Key Stage 3.

Inevitably any "real life book" that tries to stick to a syllabus will be slightly false and not necessarily the mathematics that really surrounds all of us. But this tongue-in-cheek book should give rise to potential mathematical conversations while providing valuable practice of exam style questions.

Meet the Jones family; they are a mathematical family that find maths in whatever they are doing. They are Mr and Mrs Jones, Henry Jones (aged 13), Elizabeth Jones (aged 11) and their pet dog Sheba (after "The Queen of").

Their next door neighbours are the Carlisle family; Dr and Mrs Carlisle, Charlotte Carlisle (aged 13) and Jonny Carlisle (aged 5). They have a pet cat called Balkis (from "The Butterfly that Stamped").

The Carlisle family are always trying to keep up with the Joneses.

The questions about the Jones family give you hints and tips as to how to answer the questions. If you need more help you will need to look back at your school notes.

You can get more practice on similar questions from the Carlisle family.

Each chapter has a mix of different questions so that you can practice a variety of topics at the same time and get used to answering a mixture of questions.

All questions have space for all your beautiful mathematical workings, as well as the answers.

At the end of each chapter, there is a blank page for you to write your own questions about your family. You could also use this page for writing revision notes.

CHAPTER 1
THE PARTY

Jones Questions

Mr and Mrs Jones love to entertain. Below is Mrs Jones' recipe for cheese straws.

> For 6 people:
> 370 g plain flour
> 230 g butter
> 210 g cheese

1) a) Mrs Jones has 15 people coming to the party and is wondering how much of each ingredient she will need. How much butter would she need?

Hint: You can find out how much 3 people would need by dividing by 2 and then work out how much you need for 15 people by multiplying by 5

Answer: .... g (2)

b) Mrs Jones discovers that she only has 280 g of cheese. How many people will she be able to feed?

Hint: If you know how many grams are needed to feed 1 person, you can work out how many people 280 g will feed.

Answer: .... people (2)

Mr Jones wonders how much this will cost.

2) a) Round the number of grams in the recipe for 6 people to 1 significant figure.

Hint: To round to 1 significant figure, you need to approximate the number using only the first digit.

Flour ……………………………… g   Butter ……………………………… g   Cheese ……………………………… g (2)

Price List

Flour costs 10p per 100 g
Butter costs 38p per 100 g
Cheese costs 85p per 100 g

b) Using your answer to question 2a) and rounding the prices to 1 significant figure, estimate the cost of the recipe for 6 people. Give your answer to 1 significant figure.

Answer: £ ……………………………… (2)

3) Find the actual cost for 210 g of cheese at 85p per 100 g. Give your answer in pounds and pence.

Answer: £ ……………………………… (2)

Carlisle Questions

Mrs Carlisle has decided that she is going to have a tea party for Jonny and has found a recipe for shortbread.

For 4 hungry
shortbread lovers:

120 g butter
50 g caster sugar
180 g plain flour

1) a) How many grams of butter will Mrs Carlisle need for 2 people?

Answer: .... g (1)

b) How many grams of flour will Mrs Carlisle need for 14 people?

Answer: .... g (2)

c) Mrs Carlisle discovered that she only has 315 g of plain flour.
   How many people can she make shortbread for?

Answer: .... people (2)

2) Mrs Carlisle has 480 g of butter.

a) Write 480 to 1 significant figure.

Answer: .... (1)

Butter costs 43p per 100 g.

b) Write 43p to 1 significant figure.

Answer: .... (1)

c) Using your answers to questions 2a) and 2b), approximate how much the butter cost?

Answer: .... (1)

d) How much did the butter actually cost?

Answer: £ .... (2)

Challenge Questions

1) In order to keep up with the Joneses, Mrs Carlisle decides to have a huge party and invites the whole of Jonny's school. She knows that there are fewer than 300 students in total.

She worked out earlier that she could divide the children exactly into tables of seven or eight.

a) What is the least number of children that she could have invited?

Answer: .... children (2)

b) In the end, between 100 and 150 children accepted the invitation. How many children does she have coming to the party?

Answer: .... children (3)

2) Jonny and Charlotte decide to share a whole birthday cake. Jonny's piece is 40% bigger than Charlotte's. What fraction of the cake does Jonny eat?

Answer: .... (3)

OWN QUESTIONS

NOTES

CHAPTER 2

IN THE KITCHEN

Jones Questions

This is a net of the box of Elizabeth's favourite cereal, *Cheery Cherry Cereal*, and a 3-D picture of the box.

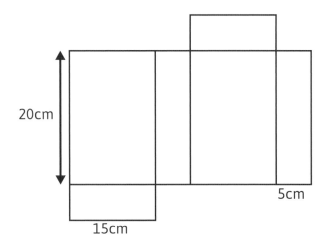

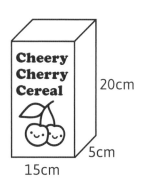

1) a) Find the volume of Elizabeth's favourite cereal box.

Hint: To find the volume of any prism, including a cuboid, you find the area of the base and multiply this by the height.

Answer: .... cm³ (2)

b) Find the surface area of the *Cheery Cherry Cereal* box.

Hint: To find the surface area you must find the area of each section of the net and then add them together.

Answer: .... cm² (2)

c) The promotion sticker on Elizabeth's box of cereal uses up 40% of the area of the front of the box. Find the area of the promotion sticker.

Hint: To find 40% of the number multiply by 0.4 or find 10% (by dividing by 10) and then multiply by 4.

Answer: .... cm² (2)

Henry's favourite pasta sauce is *Simon's Spicy Sauce.*

Simon's Spicy Sauce comes in cylindrical tins. The tin has a radius of 4 cm and a height of 10 cm.

2) a) What is the volume of the tin? Leave your answer to 3 significant figures or in terms of π.

Hint: To find the volume, find the area of the cylinder's base (π x r²) and then muliply by the height.

Answer: .... cm³ (2)

b) Henry invites the rugby team for supper and needs 3 litres of *Simon's Spicy Sauce.* How many tins will he need?

Remember: 1 litre = 1000cm³

Answer: .... tins (2)

Carlisle Questions

1) Charlotte Carlisle's favourite cereal, *Wonderful Wheat*, comes in the following box.

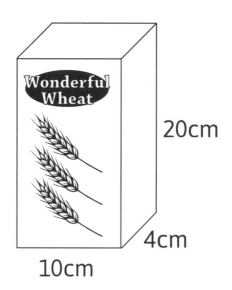

a) Find the volume of a box of *Wonderful Wheat*.

Answer: .... cm³ (2)

b) Complete the net of the box of *Wonderful Wheat*

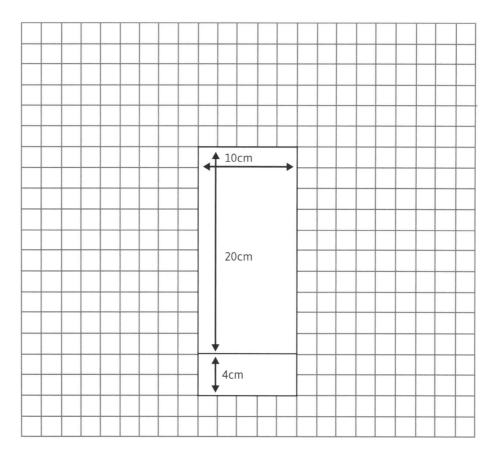

c) Find the surface area of the complete box.

Answer: .... cm² (2)

d) A special offer label on the front of the box takes up 35% of the area. What is the size of the special offer label?

Answer: .... cm² (2)

Challenge questions:

1) The tins of *Simon's Spicy Sauce* are packed in boxes of four. Each tin has a radius of 4 cm and a height of 10cm. The boxes are cuboids.

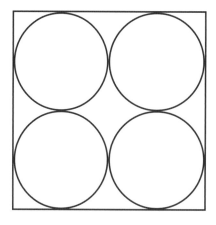

Find the volume of each box.

Answer: .... cm³ (2)

2) The manufacturer of *Simon's Spicy Sauce* is designing a new label. This label is going to be stuck over the curved face of the cylinder with no overlap.

Hint: the surface area of the curved face of a cylinder is found by multiplying the circumference of the circular base (2 x π x r) by the height of the cylinder.

Find the area of the label for Dr Carlisle.

Answer: .... cm² (2)

3) A large tin of *Simon's Spicy Sauce* contains 8 litres of sauce. $\frac{3}{4}$ of the sauce is poured into an empty cuboid container with a base area of 300 cm². If the container was $\frac{5}{6}$ filled, what is the height of the container?

Answer: .... cm (4)

OWN QUESTIONS

NOTES

CHAPTER 3

THE GARDEN

Jones Questions

The Joneses want to redesign their garden. They decide to plot the points on a co-ordinate grid.

Hint: x comes before y in the alphabet and so it also comes before y in the co-ordinates and x is a cross (corny pun intended).

1) Plot the following co-ordinates on the grid opposite and give the best names for the shapes of the flower beds. 1 unit on the grid represents 1m on the ground.

Shape 1: co-ordinates (2,3) (2,5) (4,6) (4,2)
Shape 2: co-ordinates (8,2) (11,2) (11,6)
Shape 3: co-ordinates (7,14) (10,14) (10,12) (7,12)

2) The Joneses are going to redesign their garden, on the grid opposite:

a) Reflect shape 1 in the line x = 5 and label it 1b.

Hint: The equation of a line is the rule that all the co-ordinates on that line obey. So on y = 7 all the y co-ordinates are 7. If you are unsure what the line looks like then think of some co-ordinates with the y co-ordinates equal to 7 e.g. (2, 7) or (4, 7) and then join them.

b) Translate shape 2 right 1 unit and up 4 units and label the new shape 2b.

c) Enlarge shape 3 with a scale factor of 2 and a centre of enlargement (12,15) and label it 3b.

Hint: Remember to measure the distance of the enlargement for each point from the centre of enlargement. The flower bed will move as well as grow.

3) Find the total area of the new flower beds.

Answer: .... (2)

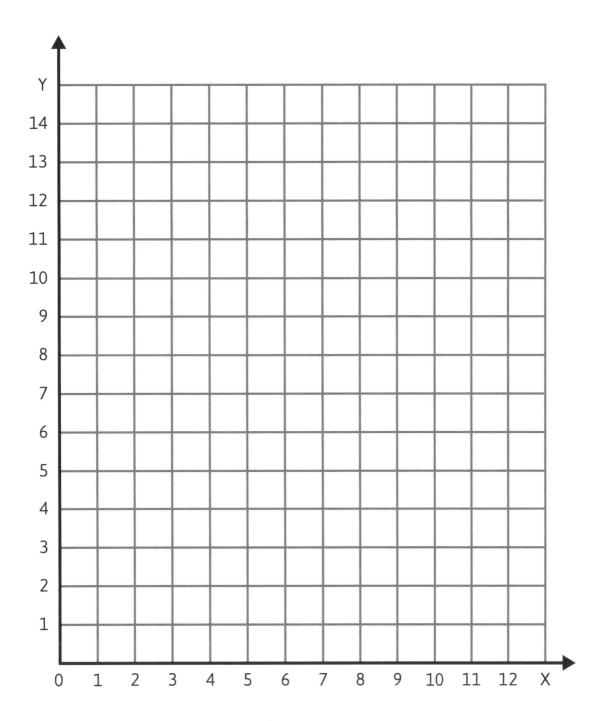

Answer:

Shape 1 is .... (2)

Shape 2 is .... (2)

Shape 3 is .... (2)

4) Henry and Elizabeth want to turn the pond into a sunken trampoline and need to dig out a cylinder of earth.

a) On the grid draw a circle of radius 2 units and a centre at (3, 13)

b) They want to dig 1.5 m down. How much soil will they need to remove?

Answer: .... m³ (2)

c) A skip holds 6 m³. How many skips will they need to hire to remove the soil?

Answer: .... (2)

d) To hire a skip it costs £180. What is the cost of hiring the skips?

Answer: £ .... (2)

Carlisle Questions

Dr and Mrs Carlisle want to redesign their garden now they see how beautiful the Jones' garden is.

1) They drew a scale drawing of their existing garden using the scale 1 : 100.

   a) Flower bed 1 is a rectangle 2 m by 3 m. Give the dimensions and units that would be used on the diagram

   Answer: .... (1)

   b) If they enlarge their flowerbed by a scale factor of 3, what will the area of the new flower bed be?

   Answer: .... m² (2)

   c) Mrs Carlisle decides to buy enough top soil to cover the new flower bed at a depth of 50 cm. What volume of top soil will she need?

   Answer: .... m³ (2)

Blended top soil is sold at £24.00 per metric tonne (1,000kg).

Mrs Carlisle knows that 1 tonne of blended top soil is approximately 1 cubic metre.

d) Using your answer from question 1c) how many metric tonnes does Mrs Carlisle need?

Answer: .... metric tonnes (1)

e) Calculate the cost of putting new top soil on her flower bed?

Answer: £ .... (2)

2) Charlotte wants to plant some spring bulbs.

The ratio of narcissi to crocuses to tulips is 4 : 5 : 3.

a) If there were 12 narcissi, how many crocuses are there?

Answer: .... crocus (2)

b) If there are 6 more crocuses than tulips. How many bulbs are there in the box?

Answer: .... bulbs (2)

c) A box of bulbs costs £5.97. What is the cost of 4 boxes of bulbs?

Answer: £ .... (2)

3) The Carlisles are very pleased with their large garden that is 60 m by 80 m.

Draw a scale drawing of the Carlisles' garden:

It includes some of the following:

    a) A rectangular pond which is 4 m by 2 m   (2)

    b) A square playground which is 3 m by 3 m   (2)

    c) A vegetable patch which is 1 m by 3 m   (3)

    d) A shed which is 2 m by 3 m   (2)

    e) A grass area which is at least 200 m²   (2)

    f) A right angled triangular flower bed with an area of 8 m² and a base of 4 m   (1)

    g) A round pond with a radius of 2 m   (2)

    h) A trapezoidal flower bed with a distance of 3m between the parallel sides   (1)

There must be paths 1 m wide from the front of the garden to each of the attractions in the garden.

Be as imaginative as you like...

(A suggested scale would be 1 : 400, meaning that 1 cm on the picture would be 4 m in real life)

The Carlisles' Garden

The Carlisles' Garden

Challenge question

1) Dr Carlisle was determined that two of his paths should meet at right angles. He does not have a protractor large enough to measure the angle in his garden but he did have a piece of rope that was 12 m long. He tells his wife that he can use this rope to guarantee that the paths are at right angles to each other.

How can he do it? (3)

Hint: Think of an ancient Greek who was born on the Island of Samos.

OWN QUESTIONS

NOTES

CHAPTER 4

THE JOURNEY TO SCHOOL

Jones Questions

On the way to school Mrs Jones always likes to spot number plates. She likes to think about the mathematician Srinivasa Ramanujan*.

Mrs Jones likes to look at number plates and find the product of prime factors of each of the number plates she sees.

G 63 TCE   LB 42   H 75 TBR

Hint: You can use a "cherry tree" or ladder method here. Index form means that 2 x 2 x 2 is written 2^3

1) Express the following numbers as products of their primes, in index form.

   a) 63

   Answer: .... (2)

   b) 42

   Answer: .... (2)

* Footnote: While Srinivasa was in hospital another mathematician, Hardy, came to visit him. Not knowing what to say he made a comment about his taxi which was number 1,729. Srinivasa immediately said 1,729 is a fascinating number. It is the smallest number that can be expressed as the sum of two cubes in two different ways. Can you work out how?

Hint: Venn diagrams can help when finding the Highest Common Factor (usually a smallish number) or the Lowest Common Multiple (a bigger number).

c) Find the lowest common multiple of 63 and 42.

Answer: .... (2)

d) Find the highest common factor of 63 and 42

Answer: .... (2)

2) a) Express 75 as a product of its primes, in index form.

Answer: .... (2)

b) What is the smallest integer value needed to multiply 75 by to make a perfect square.

Answer: .... (2)

Mr Jones values punctuality as "the prerogative of princes" and therefore likes to make sure that he can vary his speed and arrive exactly on time.

3) It is 9 km to school and it normally takes him 12 minutes.

Hint: Remember to convert the units first. 12 min = 0.2 of an hour (12 ÷ 60).

a) What is his average speed?

Answer: .... km per hour (2)

b) How fast is this in m/s ?

Hint: Do this in two stages; convert to km per second and then m per second.

Answer: .... m/s (2)

Carlisle questions

Dr Carlisle drives his children to school most mornings. He spots a number plate with the number 60 on it.

1) a) Write the number 60 as a product of its primes, giving your answer in index form.

Answer: .... (2)

b) What is the smallest integer which 60 can be multiplied by to give a square number?

Answer: .... (1)

c) The number 54 can be written as 2×3^3. Hence, or otherwise, find the Highest Common Factor of 54 and 60.

Answer .... (1)

d) Jonny seems to find himself in detention at school once every 60 days. Charlotte finds herself in detention every 54 days. They were both in detention on the first day of term. How many days will it be before they are both in detention again?

Answer: .... (2)

2) Jonny sees a number plate that has a 2 digit odd number on it.

He gives the following clue:

"The sum of the digits is 9; the product of the digits is 18"

What is the number Jonny has seen ?

Answer: .... (2)

3) Charlotte sees a number plate that has a 2 digit number on it.

She gives the following clue:

"It is the only 2 digit square number that is also a cube number."

What is the number Charlotte has seen?

Answer: .... (1)

4) Mrs Carlisle sees a number plate with a 3 digit number on it.

She gives the following clue:

"If you multiply the digits together you get a prime number"

Everyone groans as says that it is not possible but she is correct. List all the possible numbers that she could have seen.

Answer: .... (3)

5) The Carlisles live 20 km away from the school that Jonny and Charlotte attend.

Normally Dr Carlisle drives to school in 25 minutes.

a) What is his average speed for the journey?

Answer: .... km/h (2)

b) On Monday he is stuck in a traffic jam for the first 10 minutes of the journey and only travels 4 km. At what speed must he now travel so he arrives at school on time?

Answer: .... km/h (2)

c) What is this speed in metres per second?

Answer: .... m/s (2)

Challenge question

1) Mrs Carlisle travels to her parents each weekend. She normally averages 60 km/h.

Last weekend on the return journey (on exactly the same route) she took 5 hours.
The first third of the journey took 1 hour. She then slowed down and travelled at an average speed of 40 km/h for the rest of the way home.

a) How far away do Mrs Carlisle's parents live?

Answer: .... km (3)

b) How long does it normally take her?

Answer: .... hours (3)

OWN QUESTIONS

NOTES

CHAPTER 5

AT SCHOOL

Jones Questions

The school art department organised a competition where students were all asked to design a simple puzzle made of square pieces.

Henry designed the puzzle below, featuring the family's pet dog, Sheba.

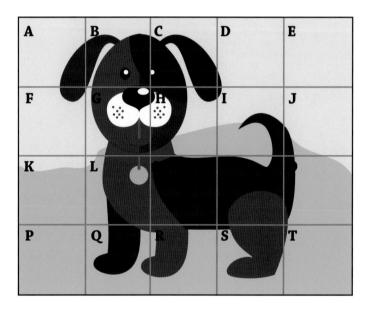

1) a) Find the percentage of the squares with some of Sheba on them.

Answer: .... % (2)

b) Find the fraction, in it's simplest form (lowest terms) of squares with some land and Sheba.

Answer: .... (2)

c) Find the ratio of the number of pieces with some sky to the number of pieces without sky.

Answer: .... : .... (2)

2) Elizabeth arranged the numbers of the squares into the correct regions of the **Carroll diagram** below.

No Sheba		
Sheba		
	Sky	No Sky

Sheba knocked all the squares onto the floor, where they all landed upside down!

a) What is the probability of picking up a piece with sky in it?

Tip: Write your probability answers as fractions

Answer: .... (2)

b) What is the probability of picking up a piece with Sheba on it?

Answer: .... (2)

c) Given that the piece has Sheba on it, what is the probability that there is also some sky?

Note: This is now only out of the pieces that have Sheba, not out of the whole puzzle.

Answer: .... (2)

d) Find the probability of picking up a piece with one of the letters of Sheba's name on it? Write your answer in its simplest form (lowest terms).

Answer: .... (2)

45

Carlisle Questions

1) Charlotte's puzzle, shown below, features the family's pet cat, Balkis.

a) What is the percentage of squares with some tree on them?

Answer: .... % (2)

b) What is the fraction, in its simplest form (lowest terms), of squares with some sky but no Balkis

Answer: .... (2)

c) What is the ratio of the number of pieces with some sky to the number of pieces without sky? Give your answer in its simplest form (lowest terms).

Answer: .... (2)

d) Write the numbers of the squares in the correct regions of the Carrol diagram below.

No Balkis		
Balkis		
	Sky	No Sky

2) Balkis knocked all the squares onto the floor, where they all landed upside down!

e) What is the probability of picking up a piece with Balkis with no sky on it?

Answer: .... (2)

f) Given that the piece has no Balkis, what is the probability that the square has sky on it?

Answer: .... (2)

g) What is the probability of picking up a piece with some tree on it?

Answer: .... (2)

2) In geography, Charlotte is studying the formation of snowflakes*.

a) A snowflake can take 1 hour to travel to the ground. They are normally formed in clouds at 3 km above the earth.

How fast are they travelling in m/s ?

Answer: .... m/s (3)

b) How fast is this compared to your normal walking pace?

*Each snowflake has a hexagonal structure because the molecules in ice crystals join to one another in a **hexagonal** structure, an arrangement which allows water molecules, each with one oxygen atom and two hydrogen atoms, to form together in the most efficient way. No two snowflakes are ever the same.

Challenge Questions

1) Last year at sports day Jonny competed in the 50 m egg and spoon race. He completed the race in 20 seconds.

a) How fast is this in m/s?

Answer: .... m/s (3)

His friend Henry ran the egg and spoon race at an average of 2.4 m/s.

b) How long did it take him to finish? Give your answer to the nearest second.

Answer: .... seconds (2)

c) Who won and by how many metres?

Answer: .... by .... metres (3)

OWN QUESTIONS

NOTES

CHAPTER 6
THE DAY TRIP

Jones Questions

While the Joneses were in London they took some photos of famous landmarks and realised that many of them had mathematical shapes that tessellated.

Below is a photo of the roof of the British Museum.

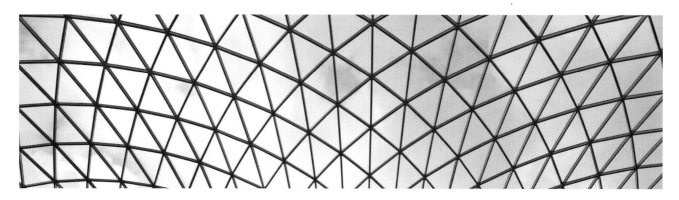

A *tessellation* is a pattern made of identical shapes:

- the shapes must fit together without any gaps
- the shapes should not overlap
- the interior angle must divide evenly into 360^0.

Hint: The **interior angle** of the polygon is equal to 180^0 minus the exterior angle.

Hint: To find the **exterior angle** of any regular polygon, divide 360° by the number of sides. To find the **number of sides** of any regular polygon, divide 360^0 by the exterior angle.

1) ABCDE is part of a regular polygon. Angle BCD is 150°.

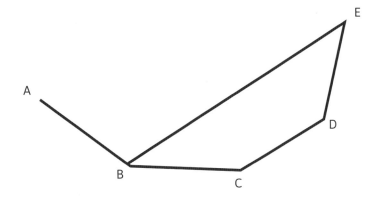

Hint: Always write on the diagram any angle you have found; it helps!

a) How many sides does the complete polygon have?

Answer: .... sides (2)

b) Find the angles

Hint: Angle CDE is the angle (smaller than 180°) that is formed when you run your finger from C to D to E.

i) Angle CDE = ...(1)
ii) Angle CBE = Angle DEB = ... (1)
iii) Angle ABE = ...(1)

c) Name the polygon enclosed by the BCDE

Answer: .... (2)

Carlisle Questions

The Carlisles took a day trip to visit the Eden project whilst they were in Cornwall. They noticed that the roof of one of the bio-domes looked like this and seemed to be made up of regular hexagons.

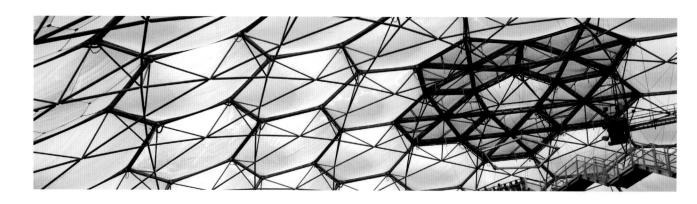

1) a) Find the exterior angle of a regular hexagon

Answer: .... (1)

b) The figure below is a hexagon where "O" is the centre.
Calculate the sizes of the following angles:

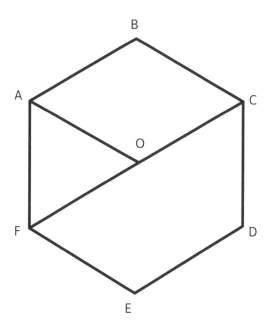

Angle AOC = ... (1)

Angle OCD = ... (1)

Angle AFO = ... (1)

c) Will a regular hexagon tessellate? Circle the correct answer and give a reason.

Answer: yes / no because ....
.... (2)

2) When Jonny cut open his apple he saw that he could image there was a pentagon around the pips.
a) Would a regular pentagon tessellate? Circle the correct answer and give a reason.

Answer: yes / no because ....

.... (2)

b) A regular pentagon is joined to a regular octagon.

Not to scale

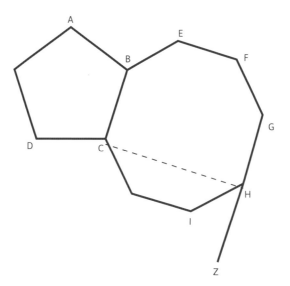

Find the following angles.

Angle ABC .... (1)

Angle CBE .... (1)

Angle IHZ .... (1)

Angle CHI .... (2)

Angle ABE .... (2)

3) Looking at the photo of the scaffolding Charlotte realised that once she knew one angle she could work out the others using parallel line theories.

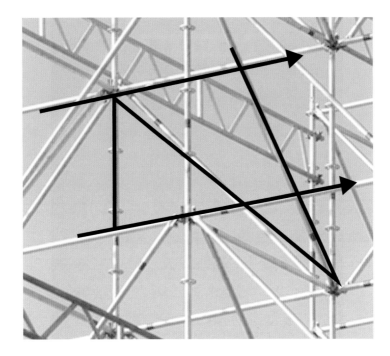

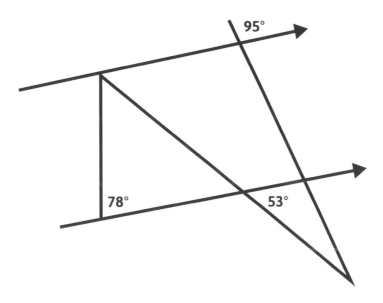

Fill in the rest of the angles for Charlotte.

4) At elevenses, the Carlisle family decide to visit the exclusive CAFCOFF cafe near Tower Bridge. Below is the menu at CAFCOFF.

MENU	
CAFCOFF special	£3.85
Hot chocolate	£3.47
World tea	£2.63
Biscuit	£3.78
Waffle	£5.89

a) Jonny orders a hot chocolate and a waffle. How much does this cost?

Answer: £ .... (2)

b) Dr Carlisle orders a CAFCOFF special and a biscuit. How much does this cost?

Answer: £ .... (2)

c) How much more do Jonny's items cost than Dr Carlisle's?

Answer: £ .... (2)

d) At the end of the meal Dr Carlisle decided to add an extra 15% as a tip to the bill. How much does the bill come to altogether?

Answer: £ .... (3)

Challenge questions (Try to do these without a calculator)

1) Jonny has £45 to spend in the shops in London. He needs to buy presents for his parents, his sister, his Granny and Grandpa and the four members of the Jones family.

a) What is the average (mean) amount that he can spend on each person?

Answer: .... (2)

He spent £4.95 on his mother, £3.65 on his father and £7.80 on his sister.

b) How much has he spent altogether?

Answer: .... (2)

c) How much does he have left?

Answer: .... (2)

d) Given that he spent all his money and he spent $\frac{2}{5}$ of the remaining amount on the Carlisle family, how much has he spent on his Granny and Grandpa?

Answer: .... (3)

OWN QUESTIONS

NOTES

CHAPTER 7

THE SKIING HOLIDAY

Jones Questions

The Joneses love to ski in Val d'Isere.
This year the exchange rate is £1.00 (Pounds) = €1.4 (Euros)

1) a) How many euros would Mr Jones get if he exchanged £750?

Hint: When you exchange from Pounds to Euros you need to multiply by the exchange rate.

Answer: .... Euros (2)

b) Mrs Jones' hot chocolate in the mountain café cost €8.05. How much is this in pounds?

Hint: When you convert back you have to do the opposite. So divide by the exchange rate.

Answer: £ .... (2)

c) The exchange rate is now £1.00 = €1.4
   Last year the exchange rate was £1.00 = €1.2

What is the percentage increase from last year to this year?

Hint: Useful formula to learn:

$$Percentage = \frac{difference}{original} \times 100$$

Answer: .... (2)

d) Elizabeth said this was a good thing as you got more euros for each pound

Is she correct and why? ....

.... (2)

e) Why does the exchange rate matter? ....

.... (2)

2) Henry noticed that the Blue, Red and Black runs were in the ratio 5 : 4 : 2

Elizabeth decides that she wants to keep the same ratio when she is skiing the runs. She wants to do 20 Blue runs.

a) How many runs in total would she have to do?

Answer: .... runs (1)

b) If Elizabeth skied for 5 and a half hours, what is the average length of time each run would take.

Answer: .... mins (2)

b) (i) Do you think this is a reasonable number to do?

Answer: ....

....

c) Mr Jones wants to ski Blue and Black runs. If he wants to ski 15 more Blue than Black runs how many of each will he need to ski?

Blue runs: .... (1)

Black runs: .... (1)

Carlisle Questions

The Carlisles decided to wait until the Easter holidays before they went skiing. Jonny was then just old enough to join the ski school.

1) When the Carlisles went skiing, the exchange rate was £1.00 (Pounds) = €1.55 (Euros)

They thought that by waiting they had a much better deal.

a) When Dr Carlisle exchanged his £750 how many euros did he get?

Answer: € .... (2)

b) Mrs Carlisle's hot chocolate, with rum, cost €10.35. What was the equivalent cost in pounds?

Answer: £ .... (2)

c) Charlotte saw a new glow ski suit in one of the shops costing €364.65, how much would that be in pounds?

Answer: £ .... (2)

d) Since the Joneses went skiing, the exchange rate has risen from £1 = €1.4 to £1 = €1.65. What is the percentage increase in the exchange rate?

Answer: .... % (2)

2) When the Carlisles went skiing the ratio of the Blue, Red and Black runs had changed. More runs were open and the ratio was now 3 : 2 : 1

a) If there were 33 Blues runs open. How many Red runs were open?

Answer: .... (2)

b) The next day it was snowing really hard and so there were only 18 Red runs open. The ratio of the runs remained the same. How many runs were open in total?

Answer: .... (2)

c) The day after, there was 16 more Blue runs than Black runs. How many Red runs were open?

Answer: .... (2)

d) Dr Carlisle had an app on his phone that told him how many kilometres he had travelled. He completed 35 runs in 9 hours. On average how many minutes did it take him to complete each run?

Answer: .... (2)

Hint: Be careful when converting from minutes to minutes and seconds (0.4 of a minute is 24 seconds).

Challenge question

The number of runs completed by Jonny to Charlotte on Monday was 2 : 1

On Tuesday, Jonny skied 9 more runs than Charlotte. However, Jonny skied $\frac{3}{10}$ fewer runs than he had on Monday and Charlotte skied $\frac{1}{5}$ fewer runs than she had on Monday.

How many runs did Jonny ski on Monday?

Answer: .... (4)

OWN QUESTIONS

NOTES

CHAPTER 8

THE SAILING HOLIDAY

Jones Questions

While sailing in the Maldives the Jones family sail for 12 km from port A on a bearing of 060°.

At buoy B, they change to a bearing of 150° for 5km until they reach point C.

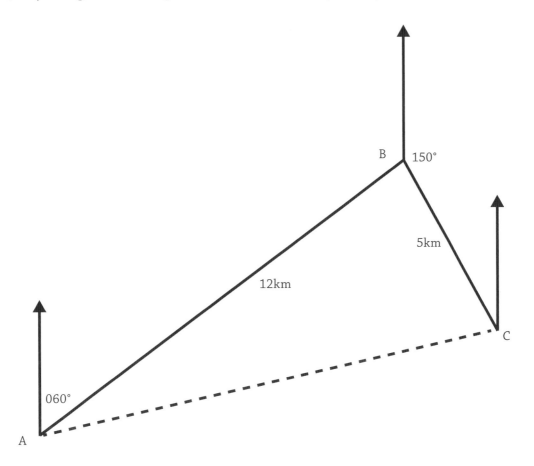

They want to know what bearing they need and how far they need to travel to get back to port A from point C.

1) a) Draw a scale drawing of their journey. (2)

Hint: You should draw this accurately using a compass and a protractor. The compass is used for finding directon accurately and drawing circles and arcs and the protractor for angles.
Remember that bearings are always 3 digits and measured clockwise from the North line.

b) What is the bearing of A from B?

Answer: .... (1)

c) Find the bearing from C to A.

Answer: .... (2)

d) The scale of the map is 1 : 100000. How many km would 7 cm on the map represent?

Hint: 1 : 1000 means 1 cm on the map is 1000 cm on the land.
1 cm = 10 m

Answer: .... km (1)

e) Find the distance, in km, from C to A.

Answer: .... km (1)

Challenge Question

Work out the distance (in km) from C to A without using the scale diagram. (2)

Carlisle Questions

1) The Carlisles are on a sailing holiday in Sardinia. The map shows the position of a restaurant and the Carlisles' yacht.

The map has been drawn to a scale of 1 : 50000

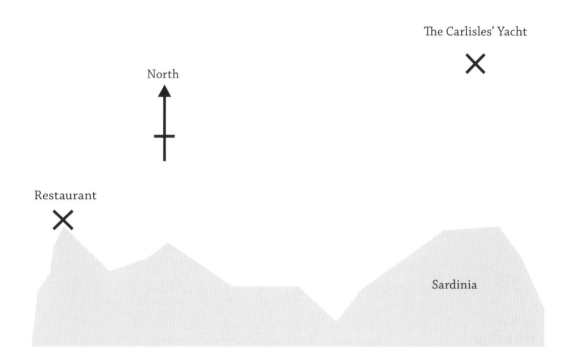

a) What is the distance of the yacht from the restaurant? Give your answer in kilometres.

Answer: .... km (2)

b) What is the bearing of the yacht from the restaurant?

Answer: .... ° (2)

c) What is the bearing of the restaurant from the yacht?

Answer: .... ° (2)

d) A buoy is 3.5 km from the restaurant and on a bearing of 280⁰ from their yacht. Mark the position of the buoy with an X on the map.

Hint: You should be using your compass to plot the distance from the restaurant.

Answer (2)

OWN QUESTIONS

NOTES

CHAPTER 9

THE FRENCH GITE

Jones Questions

The Joneses do like their French holiday each year.

1) a) The ferry journey cost £19.00 for the car and two passengers, plus £8.77 for each additional passenger, exclusive of V. A.T

What was the cost before V.A.T was added?

Answer: £ .... (2)

b) V.A.T is charged at 20%.

When you are charged V.A.T you have to add 20%. Multiply by 1.2 to find the final amount.

How much do they have to pay in total, including V.A.T?

Answer: £ .... (2)

2) Sheba, their dog does not like to travel so the Joneses leave Sheba with their neighbours, the Carlisle family. Sheba eats $\frac{3}{5}$ of a packet of *Minced Morsals* a day.

Mrs Jones wonders how many packets she would need to leave with the Carlisles if they were going away for 15 days.

Hint: You can view this as a ratio question ....
of a packet in 1 day therefore 3 packets in 5 days ......

Answer: .... packets (2)

3) The water skiing lessons cost €10 for $\frac{2}{3}$ of an hour. Mr Jones has decided that he will pay €150 for lessons and wants to know how many hours he will get?

Answer: .... hours (2)

Carlisle Questions

The Carlisles decided to fly to the south of France for their holiday.

1) a) They travelled by aeroplane with *SUPER-SPEEDY*. Tickets for each member of the family were £94.55 without taxes. How much did they pay for all four of the family?

£ .... (2)

   b) The airport added 22% for duty and tax to the ticket. How much did they have to pay after tax was added?

£ .... (2)

2) Much to everyones sadness, when the Carlisles go on holiday to their house in the south of France, they have to leave Balkis, the cat, at home with the Joneses.

Balkis eats $\frac{3}{4}$ of a tin of KITKIT cat food a day.

a) The Carlisles are going on holiday for 12 days. How many tins of KITKIT cat food should they leave?

Answer: .... tins (2)

b) Last year they left 18 tins of KITKIT cat food but Balkis only ate $\frac{3}{5}$ of a tin each day. How many days did the 18 tins last?

Answer: .... days (2)

c) Tins are bought in packs of 24. Last time 6 were dented. What percentage of tins bought were dented?

Answer: .... % (2)

3) The roof of the holiday house in the south of France has roofing tiles that look like the diagram below. The dimensions are on the tile.

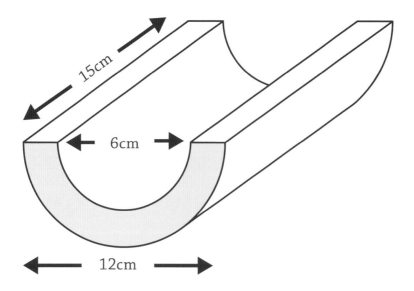

a) Find the area of a circle with a radius of 3 cm.

Answer: .... cm² (2)

b) Find the area of the shaded surface of the tile.

Answer: .... cm²(2)

c) Find the volume of each tile.

Answer: .... cm³(3)

The roofing tiles are constructed from terracotta clay. 1 tile has a mass of 0.75kg before baking. There is a 20% reduction in mass in the baking process.

d) Find the mass of a tile after baking. Give your answer in grams.

Answer: .... g(2)

The roof has 29,000 tiles.

Each tile costs 79 pence to make and is sold for £1.20.

e) How much will it cost the Carlisle family to replace the roof?

Answer: £ .... (2)

f) How much profit will the tile manufacturers make?

Answer: £ .... (1)

g) What is the percentage profit made on each tile?

Answer: £ .... % (2)

h) What is the percentage profit made on the roof?

Answer: £ .... % (1)

Challenge questions:

On their trip the Carlisles visited the local market. Mrs Carlisle bought 3 apples and 4 oranges for €2.65 and Dr Carlisle bought 1 apple and 5 oranges in the same market for €2.35. How many euros did an orange and an apple each cost?

Hint: This can be done logically or using algebraic simultaneous equations.

Let the price of an apple be "a" pence and the price of an orange be "b" pence.

3a + 4b = €2.65
1a + 5b = €2.35

You can solve this problem using simultaneous equations or using logic. (3)

OWN QUESTIONS

NOTES

CHAPTER 10

BACK AT HOME RELAXING PLAYING GAMES

After all the adventures that the Joneses have had they like nothing better than to sit down with some algebra games.

Here are three of their favourites.

Hint: Equivalent expressions in algebra are the same even though they look different. If you substitute a number into the expression you will get the same answer.

Algebra pairs

Match the equivalent expressions.

$3(x + 4)$	$5 + 2(x+3)$	$8x^2 + 24x$	$3x - 4 + 2x + 5$
$9x^2$	$6 - 2(x - 4)$	$(3x)^2$	$4x$
$2x^2(x - 1)$	$14 - 2x$	$3x + 4x^2$	$3x + 12$
$3x - 4x + 5x$	$6 + 2(x+1)$	$11 + 2x$	$7x + 21$
$x - 7 + 4x - 2$	$8 + 2x$	$- 6x$	$7(x + 3)$
$x + x^2 + 2x$	$5x + 1$	$3x(5x^2 + 4)$	$2x - 8x$
$4x^3 - 2x^2$	$8x(x + 3)$	$15x^3 + 12x$	$5x - 9$

Equation ladder

You will need a counter or small object each and a single die.

Roll the die and move forward the number of spaces indicated. Solve the equation and then move forwards or backwards the value of the answer. If your answer is correct you are allowed a second go (but not a third). If you are incorrect, it is the end of your turn.

EQUATION GAME

43 $\frac{2x}{3}=4$	44 $\frac{3x}{5}=3$	45 $\frac{2}{3}(x+2)=4$	46 $\frac{1}{2}(x-1)=1$	47 $\frac{3}{4}(x+2)=3$	48 $\frac{2x}{3}=4$	FINISH 49 $\frac{1}{2}(x+4)=\frac{1}{3}(x+7)$
42 $5-2(x-5)=-11$	41 $4-3(x+4)=1$	40 $3+2(x+1)=11$	39 $2+3(x+4)=17$	38 $3+2(x+6)=19$	37 $4(3-2x)=-28$	36 $5(2-x)=5$
29 $10-3x=1$	30 $25-5x=20$	31 $8-8x=16$	32 $7-3x=4-2x$	33 $4-5x=x-20$	34 $2x+6=18-2x$	35 $2(3+x)=10$
28 $7x+4=x-2$	27 $5x+1=3x+7$	26 $3x+1=x+5$	25 $x+2=2x-1$	24 $5x+3=2x$	23 $x+5=6x$	22 $3x+4=x$
15 $7x+5=40$	16 $10=3x+4$	17 $1=2x-7$	18 $17=6x-1$	19 $4x+7=15$	20 $5x+18=8$	21 $4x+5=-7$
14 $6x-7=28$	13 $4x+5=9$	12 $2x-4=6$	11 $3x-7=2$	10 $3x-4=2$	9 $4x+9=25$	8 $3x+9=6$
START 1 $x+5=7$	2 $x+6=12$	3 $3x=12$	4 $7x=28$	5 $\frac{x}{3}=2$	6 $\frac{x}{2}=4$	7 $2x+5=7$

Hint: What you do to one side you must do to the other.

Algebra race track: Substitution

You will need two different coloured dice and a counter for each player which is placed on the START square. Roll the first dice and then roll the second dice, subtract the number of the second dice from the first.

If you rolled a 3 and then a 1 you subtract the number to get the value of 2, you take the positive value of 2 and substituite it into the first equation in the first square to get 2 + 1 which equals 3. Move your counter forward this amount of squares. If your sum gives you a negative number, you move backwards instead. You then wait on that square until the other players complete their move. The winner is the first person to get around the race track first on the fourth lap.

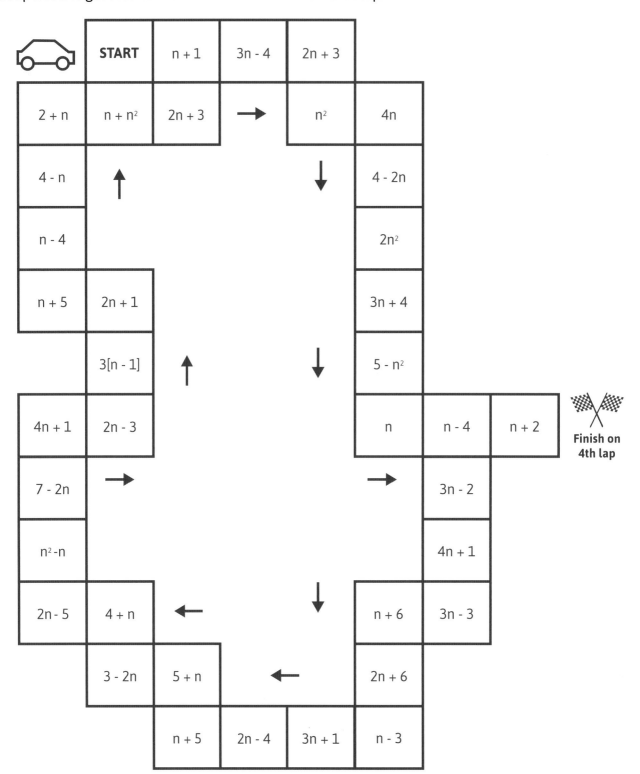

ANSWERS

Chapter 1 – The Party

Jones questions

1) a) 3 people 230 ÷ 2 = 115 g
   so 15 people 115 x 5 = 575 g

   b) 1 person 210 ÷ 6 = 35 g
   so 280g 280 ÷ 35 = 8 people

2 a) Flour 400 g, Butter 200 g, Cheese 200 g

   b) Estimate ≈ (10p x 4) + (40p x 2) + (90p x 2)
   ≈ 40 + 80 + 180
   ≈ £3.00

3) 85 + 85 + 8.5 = 198.5p this is £1.99

Carlisle questions

1) a) 120 ÷ 2 = 60 g

   b) For two people 180 ÷ 2 = 90 g
   so for 14 people 90 x 7 = 630 g

   c) For 1 person 180 ÷ 4 = 45
   315 ÷ 45 = 7 people

2) a) 480 ≈ 500 (1 sig fig)

   b) 43 ≈ 40 (1 sig fig)

   c) 5x 40 = 200p or £2.00

   d) 4.8x 43 = 206.4p or £2.06

Challenge questions

1) a) Lowest common multiple 7 x 8 = 56

   b) 2 x 56 = 112

2) If Charlotte eats x amount of cake then Jonny
   will eat 1.4x. Together they eat 2.4x = 1 whole
   cake. Therefore

   $x = \frac{1}{2.4} = \frac{5}{12}$

   So Jonny eats

   $1.4x = 1.4x \frac{5}{12} = \frac{7}{12}$

Chapter 2 – In the Kitchen

Jones Questions

1) a) 15 x 5 x 20 = 1500 cm³
   (be careful here in some questions this is an
   examiners' trap and it would be 0.0015 m³
   NOT 1.5m³ – if you get one of these questions
   ALWAYS convert to m first)

   b) Surface area
   (2 x 20 x 15) + (2 x 15 x 5) + (2 x 20 x 5)
   = 600 + 150 + 200 = 950 cm²

   c) 0.4 x 20 x 15 = 120 cm²

2) a) π x 42 x 10 = 502.65 = 503 cm³ (3 sig fig)

   b) 3000 ÷ 502.65 = 5.968 therefore 6 tins

Carlisle questions

1) a) Volume = 800 cm³

   b) Net may look like this

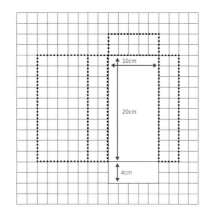

   c) (2 x 20 x 10) +(2 x 10 x 4) +(2 x 20 x 4)
   = 400 + 80 + 160 = 640 cm²

   d) 0.35 x 20 x 10 = 70 cm²

Challenge questions

1) Volume of boxes are 16 x 16 x 10 = 2560 cm³

2) Label has an area of 2 x π x 4 x 10 =
   251.327 cm² = 251 cm² (3 sig fig)

3) $\frac{5}{6}$ height x 300 = 6000

   $\frac{5}{6}$ height = 20

   height = 20 x 6 ÷ 5 = 24 cm

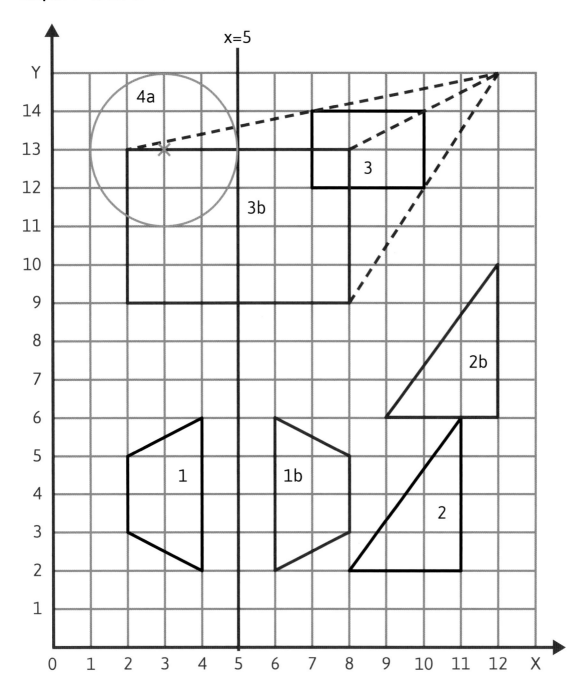

Jones Questions

1) a) Shape 1 is an isosceles trapezium, shape 2 is
   a right-angled scalene triangle, shape 3 is a
   rectangle

2) Also on grid above

3) 6 + 6 + 24 = 36 m²

4) a) See grid

   b) $\pi^2 \times 2 \times 1.5 = 6\pi = 18.8$ m³

   c) 4 skips

   d) 4 x 180 = £720

Carlisle Questions

1) a) 2 cm by 3 cm

   b) $6 \times 3^2 = 54$ m²

   c) $0.5 \times 54 = 27$ m³

   d) 27m³ = 27 metric tonnes

   e) 27 metric tonnes x £24 = £648

2) a) 15 crocus
   b) 36 bulbs
   c) £23.88

3) Everyone will draw something different but size should be relative to instructions in the question.

Challenge Questions

Use the Pythagoras' theorem. If you have a length of rope 12 m long and divide it into sections 3 m, 4 m, and 5 m. If you lay the 3 m and 4 m lengths on your paths they should be exactly 5 m apart on the hypotenuse, if your paths are at right angles.

Chapter 4 – The Journey to School

Jones Questions

1) a) $63 = 32 \times 7$

   b) $42 = 2 \times 3 \times 7$

   c) HCF = $3 \times 7 = 21$

   d) LCM = 126

2) a) $75 = 3 \times 52$

   b) 3

3) a) $9 \div 0.2 = 45$ km/h

   b) 45 km/h

   45000 m/h

   $45000 \div 3600$ m/s= 12 m/s

Carlisle Questions

1) a) $60 = 22 \times 3 \times 5$

   b) $3 \times 5 = 15$

   c) HCF = $2 \times 3 = 6$

   d) A LCM question so 540

2) 63

3) 64

4) 112, 121, 211, 113, 131, 311, 115, 151, 511, 117, 171, 711

5) a) $20 \times 60 \div 25 = 48$ km/h

   b) $16 \times 60 \div 15 = 64$ km/h

   c) $64000 \div 3600 = 17.8$ m/s (3 sig fig)

Challenge Questions

1) a) If $\frac{2}{3}$ of the journey at 40 km/h takes 4 hours, the whole journey would take 6 hours. So total distance 40 x 6 = 240 km

   b) $240 \div 60 = 4$ hours

Chapter 5 – At School

Jones Questions

1) a) $\frac{17}{20} = \frac{85}{100} = 85\%$

 b) $\frac{12}{20} = \frac{3}{5}$

 c) $13:7$

2)

No Sheba	2	1
Sheba	11	6
	Sky	No Sky

 a) $\frac{13}{20}$

 b) $\frac{17}{20}$

 c) $\frac{11}{17}$

 d) $\frac{5}{20} = \frac{1}{4}$

Carlisle Questions

1) a) $\frac{6}{24} = \frac{1}{4} = 25\%$

 b) $\frac{8}{24} = \frac{1}{3}$

 c) $20:4$ which is $5:1$

 d)

No Balkis	10	2
Balkis	8	2
	Sky	No Sky

 e) $\frac{2}{24} = \frac{1}{12}$

 f) $\frac{10}{12} = \frac{5}{6}$

 g) $\frac{7}{24}$

2 a) $\frac{3000}{60 \times 60} = \frac{5}{6}$ m/s

 b) Depending on age and fitness levels most people walk at about 5 km/h

Challenge Questions

1) a) $\frac{50}{20} = 2.5$

 b) $\frac{50}{2.4} = \frac{500}{24} = 20.8 = 21$ to nearest second

 c) Jonny won by $\frac{5}{6}$ of a second.
 Henry was therefore ($\frac{5}{6} \times 2.4 =$) 2 m behind

Chapter 6 – The Day Trip

Jones Questions
1) a) Interior angle = 150°. Therefore exterior angle = 30°. So number of sides = $\frac{360}{30}$ = 12 sides. It is a dodecagon.

   b) Angle CDE = 150°
     Angle CBE = Angle DEB = 135°
     Angle ABE = 120°

   c) Polygon BCDE is an isosceles trapezium

Carlisle Questions
1) a) $\frac{360}{6}$ = 60°

   b) Angle AOC = 120°
     Angle OCD = 60°
     Angle AFO = 60°

   c) Yes, as each interior angle is 120° and 3 x 120° = 360°

2) a) No, each interior angle is 108°, which is not a factor of 360°

   b) Angle ABC =108°
     Angle CBE = 135°
     Angle IHZ = 45°
     Angle CHI = 45°
     Angle ABE = 72° + 45° = 117°

3)

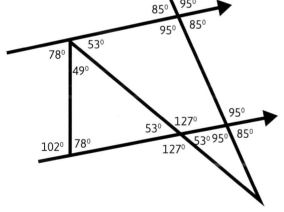

4) a) £8.06
   b) £7.63
   c) 43p
   d) 1.15 x 15.69 = £18.04

Challenge questions
1) a) 45 ÷ 9 = £5.00
   b) £16.40
   c) £28.60
   d) $\frac{3}{5}$ x 28.60 = £17.16

Chapter 7 – The Skiing Holiday

Jones Questions
1) a) 750 x 1.4 = €1050

   b) 8.05 ÷ 1.4 = £5.75

   c) $\frac{1.4-1.2}{1.2}$ x 100 = 16.7% (3 sig fig)

   d) Yes, when you are on holiday you want the exchange rate to be as high as possible so that you get more of the new currency.

   e) If you are importing or exporting goods to another country you need to check the exchange rate to see how much things cost in your own currency.

2) a) 20 : 16 : 8 so a total of 20 + 16 + 8 = 44 runs

   b) (5 x 60) + 30 = 330 min, 330 ÷ 44 = 7.5 min per run.

   bi) This is extremely unlikely if you think about times waiting for the lifts and the lifts themselves, as well as the getting down the slope.

   c) Originally a "gap" of 3 between Black and Blue ratio. Now want 15 so multiply each by 5.
     5 x 5 = 25 (blue) and 2 x 5 = 10 (black)

Carlisle Questions
1) a) €1,162.5

   b) £6.68

   c) €364.65 ÷ 1.55 = £235.26

   d) $\frac{1.65-1.4}{1.4}$ x 100 = 17.9%

2) a) 2 x 11 = 22

   b) 27 : 18 : 9 so a total of 54 runs open

   c) 2 x 8 = 16 runs

   d) 540 ÷ 35 = 15 minutes 26 seconds (He didn't have a long lunch in the sun!)

Challenge Question
1) Monday ratio 2x : 1x
   Tuesday 1.4x – 0.8x = 9
   therefore x = 15 so 2x = 30 runs on Monday

Chapter 8 – The Sailing Holiday

Jones Questions

1) a) You should have drawn a scale drawing that looks very similar to the one in the book.

   b) 240°

   c) Between 260° and 265° is acceptable. Actual answer is 262.6°

   d) 7 cm on the map would be 7 km in real life.

   e) 13 km

Challenge question

13 km using Pythagoras' theorem and parallel line angle rules.

Carlisle Questions

1) a) 6 km

   b) Anything between 68° and 72°

   c) Anything between 248° and 252°

   d) See diagram. Pupils should draw a circle of 7 cm with its centre at the restaurant and then a line with a bearing of 280° from the yacht. Where these two intersect is the buoy. These lines should always be left for inspection.

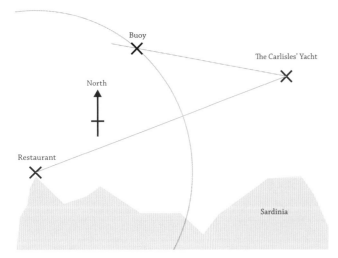

Chapter 9 – The French Gite

Jones Questions

1) a) £36.54

   b) £43.85

2) 3 packets / 5 days therefore 9 packets in 15 days

3) €30 for 2 hours so €150 for 10 hours

Carlisle Questions

1) a) 94.55 x 4 = £378.20

   b) 1.22 x 378.20 = £461.40

2) a) 9 tins

   b) 30 days

   c) 25%

3) a) $\pi \times 3^2 = 28.3$ cm^2

   b) $(36 - 9)\pi = 84.8$ cm^2

   c) 84.8 x 15 = 1272 cm^3

   d) 750 x 0.8 = 600 g

   e) 1.2 x 2900 = £34800

   f) 120 – 79 = 41 pence
      41 x 2900 = £1189 profit

   g) $\frac{41}{79} \times 100 = 51.90\%$

   h) Exactly the same, 51.90%

Challenge Questions

Apples cost 35 cents and oranges cost 40 cents.